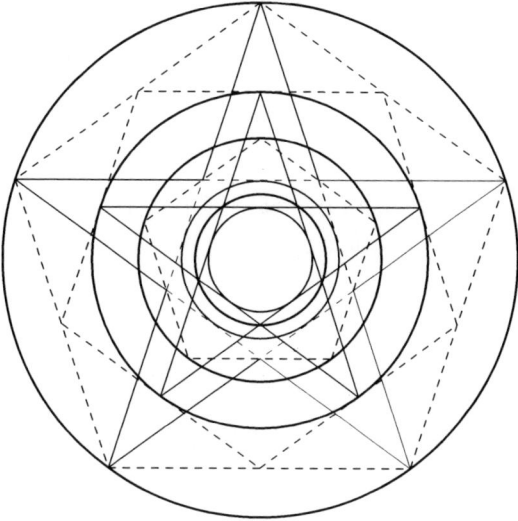

First published 1997
© Wooden Books 1997

Published by Wooden Books Ltd.
Walkmill, Cascob, Presteigne, Powys, Wales LD8 2NT

British Library Cataloguing in Publication Data
Haynes, Ofmil C. 1967-
The Harmony of the Spheres

A CIP catalogue record for this book is
available from the British Library

ISBN 0 9525862 3 1

Printed in Great Britain by
Woolnough Bookbinding Ltd,
Irthlingborough, Northants.

WOODEN
BOOKS

THE
HARMONY
OF THE
SPHERES

by

Ofmil C. Haynes

Here is a riddle
Oh riddle me how
A cat and a fiddle aquainted a Cow

A Light and another
A mirror of light
Throw day into dark
And a shadow through night

Answer this riddle
And riddle me why
These eyes from this middle
Unite in the sky

CONTENTS

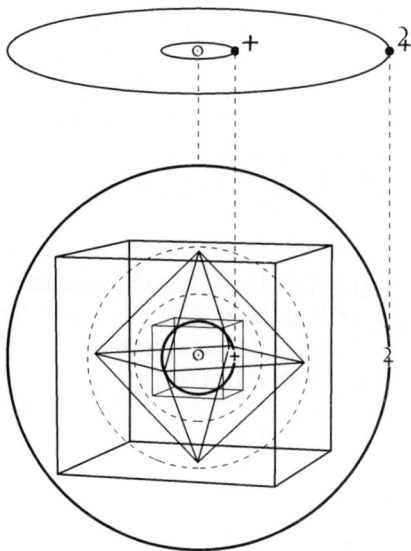

INTRODUCTION

Although this small book bears my name, only some of the research it contains is my own. The rough brew I have squeezed it from was an early Wooden Book called *A Book of Coincidence*, written by John Martineau as a thesis for Keith Critchlow and John Michell. So loud and rude were my protestations when it went out of print that I was offered the unenviable task of penning its replacement. My apologies for jewels left out; I hope some of my own additions will be welcome.

My late father once yelled at me "Beauty *is* one of the rules you idiot!" - I had been surprised by the Golden Section. The Harmony of the Spheres likewise surprises and reassures. Please do chew this little book over, sleep on it, read it again and give a copy to everyone you know! And the next time you see Venus...

Ofmil C. Haynes Jnr
Boulder, Co. March 1997

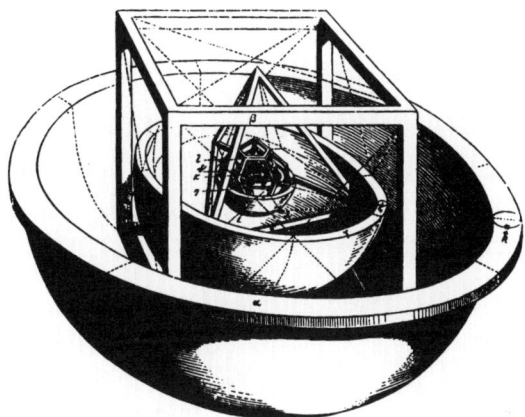

The Signs for the Heavenly Bodies

The signs for the planets in this book are shown opposite, penned by English calligrapher Mark Mills. Each can be understood as a combination of the signs for Sun, Moon and Earth.

The sign for Mars is not new and the signs for Ceres and Neptune first appeared in the last century when they were discovered. The signs for Uranus and Pluto were reworked.

Only the ancient seven heavenly bodies can be seen with the naked eye - the Sun, the Moon, Mercury, Venus, Mars, Jupiter and Saturn. Chiron and Pluto are both very small and have wobbly orbits.

The Sun is the centre of the modern Solar System - but we still live here, on Earth, and it is from this centre that we experience the heavens.

Sun

Moon

Mercury

Venus

Earth

Mars

Ceres

Jupiter

Saturn

Chiron

Uranus

Neptune

Pluto

The Solar System is a bit like a flying saucer. Four tiny inner planets whizz round the Sun quite quickly (Mercury, Venus, Earth and Mars), then there is a gap full of asteroids, the largest of which is Ceres, and then four huge outer planets (Jupiter, Saturn, Uranus and Neptune) which lumber slowly along.

All the planets go the same way round the Sun and all stay more or less on the saucer. The orbits can be pictured as circles (shown below), or spheres (page 1).

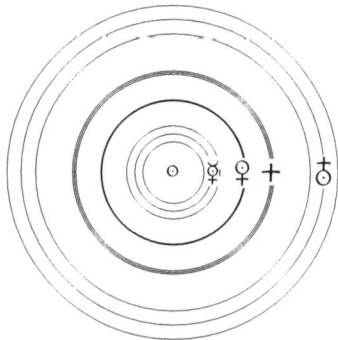

Body	Diameter	Mean Orbit	Min Max Orbits		Period
	(Miles)	*(Millions of Miles)*			*(Days)*
Mercury	3031	35.93	28.58	43.38	87.97
Venus	7521	67.24	66.78	67.78	224.7
Earth	7926	92.96	91.40	94.51	365.26
Mars	4222	141.6	128.4	154.9	687.0
Ceres	584	257.5	237.0	277.3	1680
Jupiter	89400	483.6	460.2	507.0	4332
Saturn	74900	886.7	837.3	936.1	10759
Chiron	136	1273	794.1	1757	18511
Uranus	31760	1784	1700	1868	30685
Neptune	31400	2794	2770	2818	60190
Pluto	1444	3674	2763	4586	90465
(Sun	864950)	-	-	-	-
(Moon	2160)	-	-	-	-

The Sizes of the Earth and the Moon

The Earth has a diameter of 7,920 miles (radius 3,960 miles), the Moon 2,160 miles (radius 1,080 miles). In John Michell's famous words, "The sizes of the Earth and the Moon thus make an exact 11:3 ratio".

If we draw a circle to represent the Earth (size 11), and we draw down the Moon to sit on the Earth as shown opposite (size 3), then the square round the Earth has a perimeter equal to the circumference of the circle through the centre of the Moon (with π as 22/7 both have length 44). This is fun and highly accurate.

It is interesting that the combined radii of the Earth and the Moon (3960+1080) make 5040 miles, which is 1x2x3x4x5x6x7. 7920 miles, the diameter of the Earth, then works out as 8x9x10x11.

The Europeans are making our Mile illegal!

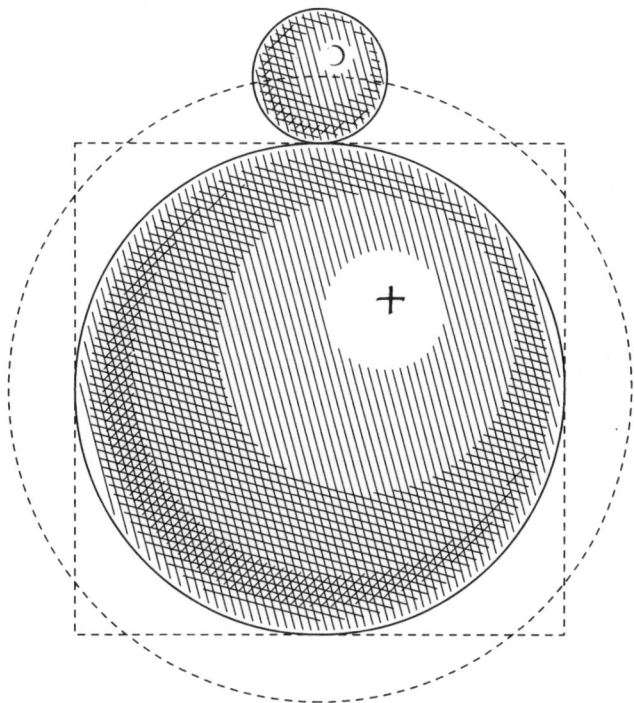

THE ORBITS OF MERCURY AND VENUS

The two innermost planets are Mercury and Venus. To draw their orbits simply put three equal circles together so that they each touch each other.

If the circle through their centres is taken to represent Mercury's mean orbit round the Sun then the circle containing the whole pattern is Venus' mean orbit. The accuracy of the solution is 99.9%.

This is an easy trick to remember. Mercury is Hermes, traditionally hermaphoditic (like a snail). Venus, Aphrodite, or the Morning and Evening Star, has the most circular orbit of all the planets, and traditionally concerns herself with Love, Beauty and Harmony.

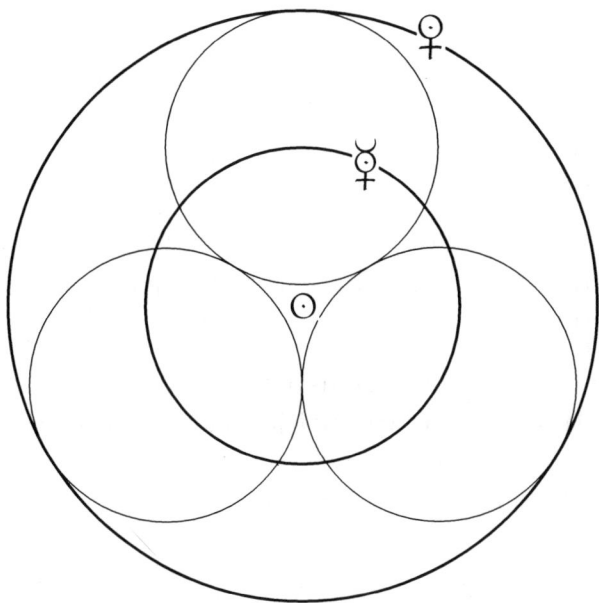

MERCURY AND THE EARTH

The diagram on the right displays a method for constructing the relative orbits of Mercury and Earth to an accuracy of 99.9%.

Two other methods of finding the proportion of the orbits of these two planets are shown below. The equivalence between them all is well worth noting.

The physical sizes of Mercury and Earth are in the same ratio to each other as are their orbits - so these diagrams could be for the relative sizes or for the relative orbits of these two planets (see page 38).

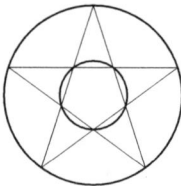

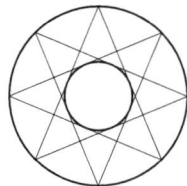

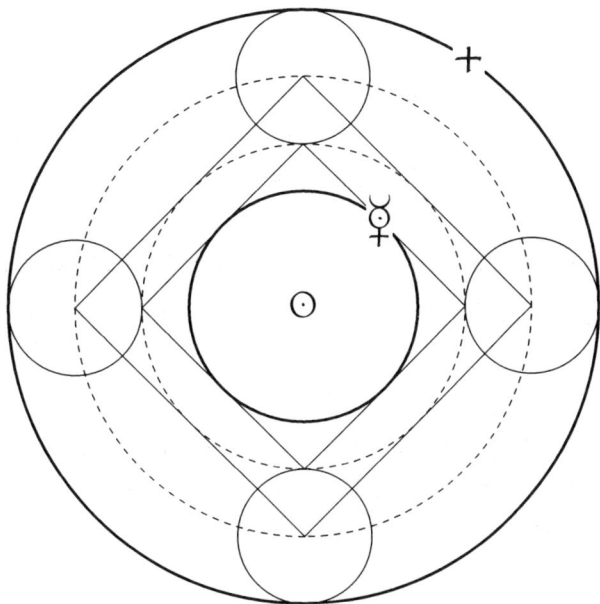

THE ORBITS OF VENUS AND THE EARTH

Venus comes closer to us than any other planet as we journey together round the Sun. To draw her orbit relative to our own simply arrange eight equal circles in a circle as shown opposite.

If the circle through the centres of the eight circles is taken to represent Venus' mean orbit then the circle enclosing the whole pattern is our own orbit. The accuracy is over 99.9%.

For fun try using three new ten pence pieces to show the Mercury-Venus proportion and eight one penny pieces for the Venus-Earth proportion.

The eight circles represent eight years.

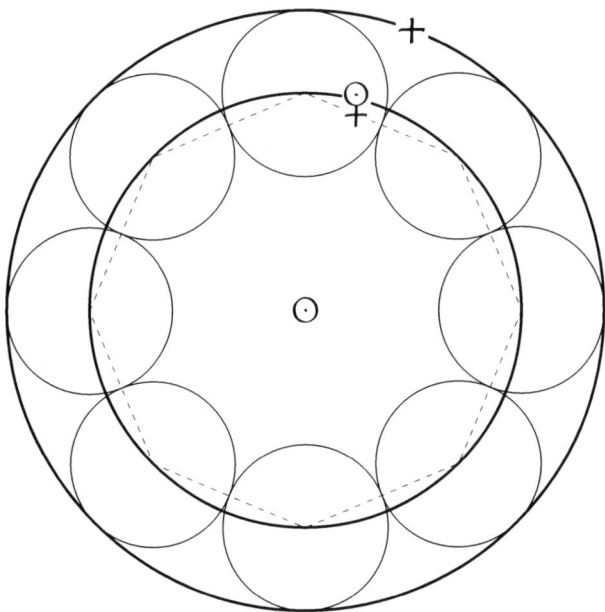

THE DANCE OF VENUS

The passage of Venus around the Earth is shown opposite. She comes in, seems to go backwards across the stars and then flies out again - and so on.

What is so beautiful is that every time Venus is closest to us, when she is between us and the Sun, she is precisely two fifths of the way round the heavens from the last place she came close.

It takes her two and a half days short of eight years to complete the five kisses. Five and eight again.

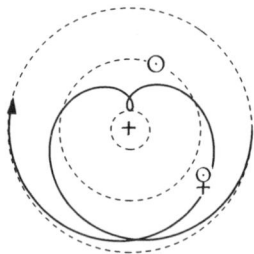

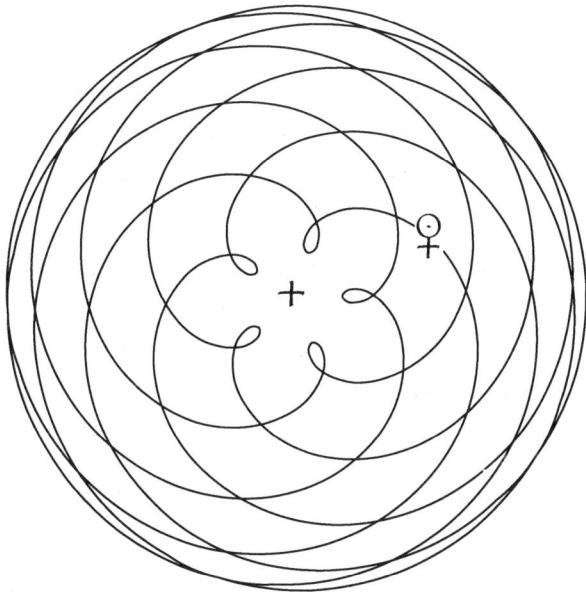

THE KISS OF VENUS

Venus goes round the Sun and the Sun goes round the Earth - that's how it looks from here. So Venus is sometimes very close to us and sometimes she is a long way away, on the far side of the Sun.

Venus' comings and goings from us form her 'body' which can be proportioned using two pentagrams with over 99.9% accuracy. The proportion is equal to the mysterious Golden Section raised to the power of four.

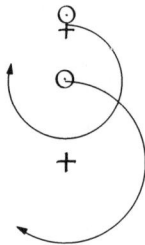

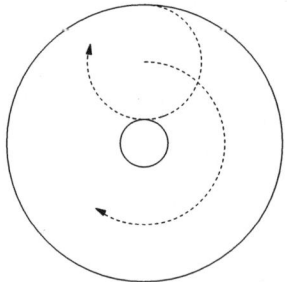

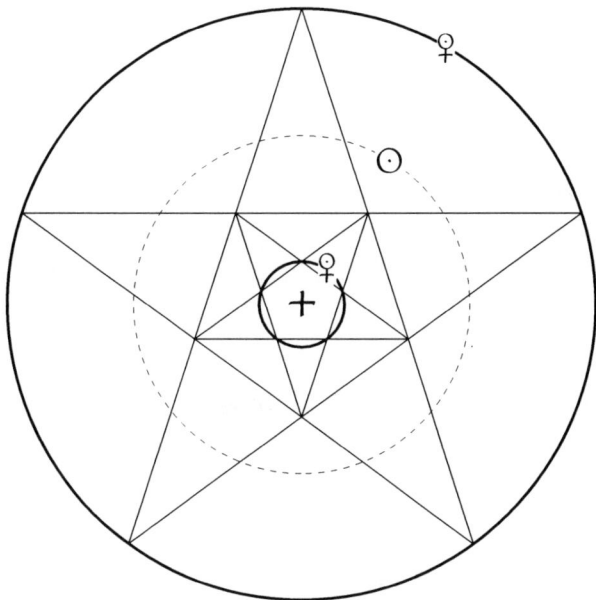

Lunar Numbers

The Moon has three colours (black, white and red) and four quarters (new, waxing, full and waning) of seven days each. Twenty-nine and a half days lie between full Moons but twenty-eight is more often used - four sevens, or 1+2+3+4+5+6+7. Four is halfway between one and seven, seven is halfway between one and thirteen and thirteen twenty-eights make 364.

Divide thirteen moonths into four quarters each (call them suits) - to make a pack of cards. Add up the pack (Jack as 11, Queen as 12 , King as 13) to make 364. Don't forget the Joker! Add one - a year and a day.

Every time the Moon comes closest to the Earth she is exactly one eighth of the way round the heavens as seen from Earth. She does this every twenty-seven and a half days, the time she takes to spin once on her axis to face the stars. The Sun takes the same number of days to spin once on its axis to face the Earth.

A human pregnancy lasts exactly nine full moons.

The Orbits of Venus and Mars

If you look at the diagram opposite you will see a geometrical solid drawn in dashed line; it is called a dodecahedron and has twelve pentagonal faces. Another way of making a dodecahedron is by arranging twenty balls as shown.

The twenty balls leave a central space which we can fill with the sphere of Venus' orbit. The whole structure can then be enclosed in the sphere of Mars' orbit with over 99.9% accuracy.

There are only five simple solids and the dodecahedron is normally placed as the fifth. In more ancient times it was linked with a mysterious fifth element of Ether.

Venus and Mars are our two neighbours - we orbit the Sun between them. Perhaps the etheric body which separates them emanates from Earth.

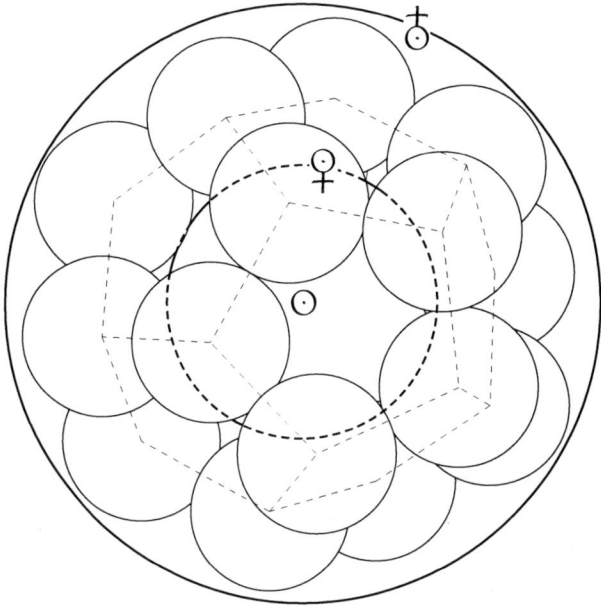

The Orbits of Earth and Mars

The dashed solid shown on the opposite page is called an icosahedron - and it is also shown as an arrangement of twelve balls.

The centres of the balls (joined by the dashed lines) lie on the sphere of Earth's orbit.

The sphere of Mars' orbit then encloses the whole structure with 99.9% accuarcy.

In the Traditions the icosahedron is associated with the fourth element of water, vital for biological life here on Earth. Turn back one page! The icosahedron and the dodecahedron are related - each of these solids is produced from the centres of the faces of the other.

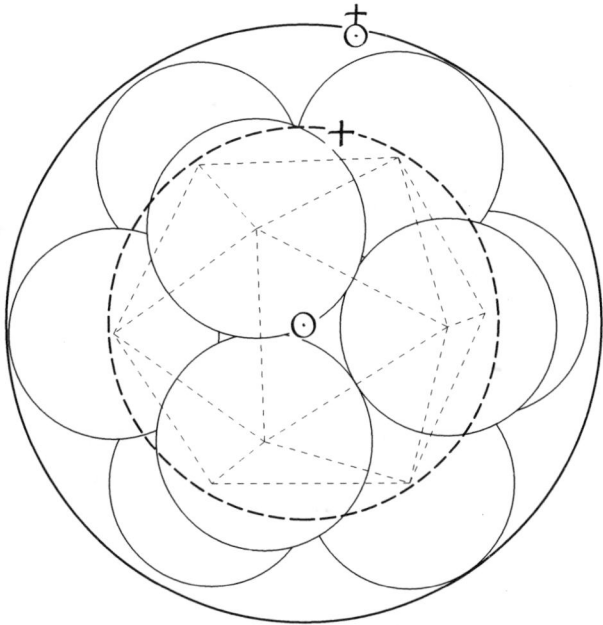

The Relationship of Venus and Mars

Venus and Mars are our two neighbouring planets. Say that you live on one of these planets and you watch the other moving relative to you over time as you both go round the Sun. The path that the other makes to and from you against the fixed stars is shown opposite in two stages, one in solid line, one dashed, each lasting roughly a year and separated by about eight years.

If their nearestness is the size of the Moon, then their furthestness is the size of the Earth. This is 99.9% accurate, hard to comprehend and quite delightful.

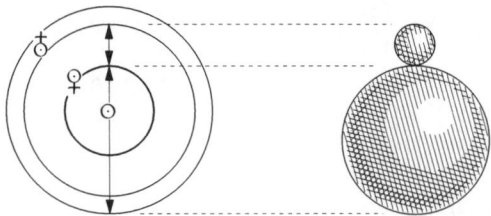

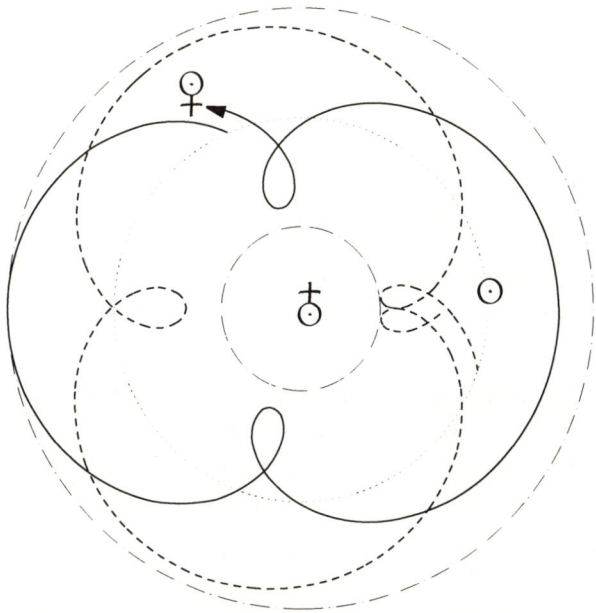

Earth and Mars again

Here's a good way to draw the orbits of Earth and Mars round the Sun:

Make a square of four circles.

Find the spacer circle left in the middle and add it to four other corners around the larger four circles.

Draw two large circles - one to the inner side of the four spacer circles, one to the outer side.

These two large circles represnt the orbits of Earth and Mars with 99.9% accuracy. This diagram is closely related to the Mars-Jupiter solution on page 33.

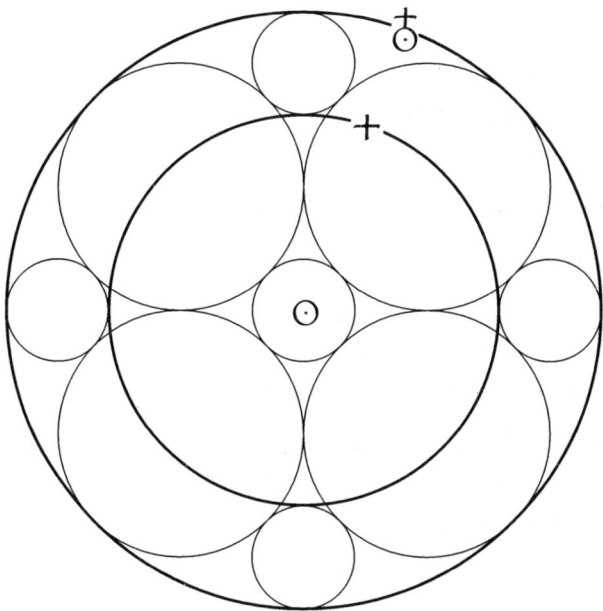

EARTH AND JUPITER

This one is really easy! Six triangles space Earth and Jupiter. A child can do it in less than five minutes and it's over 99.9% accurate. Begin by drawing a circle.

Keep the compasses the same, and, starting with the point at the top of the circle, walk round the circle ticking off the next place to put the point.

Now join up the lines as in the diagram. Earth's orbit is found at the innermost crossing radius.

Jupiter, or Zeus, used to be the King of the Gods in ancient days. The six-pointed star is often called the Seal of Solomon, or the Star of David.

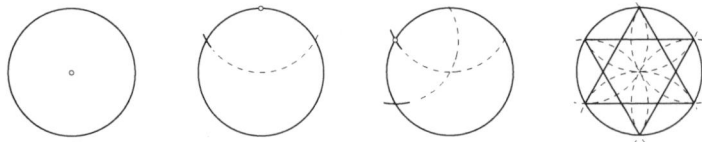

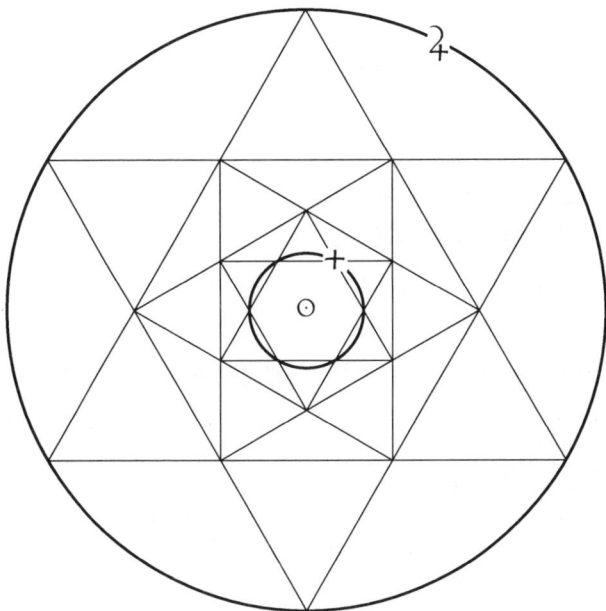

CERES

The solid shown opposite is called a tetrahedron. *Tetra* means "four" and *hedron* means "face". It is the most simple three dimensional solid possible (other than a sphere). It is shown as four balls and as the lines connecting their centres.

If the sphere through the centres of the balls is the sphere of Mars' orbit then the sphere around them all is the sphere of Ceres' orbit to an accuracy of 99.9%.

Ceres is the largest asteroid in the asteroid belt between Mars and Jupiter. She is about the size of the British Isles. Her relationship with Venus is shown below left and her orbit out to Jupiter below right.

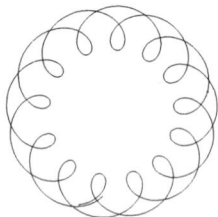

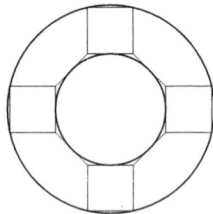

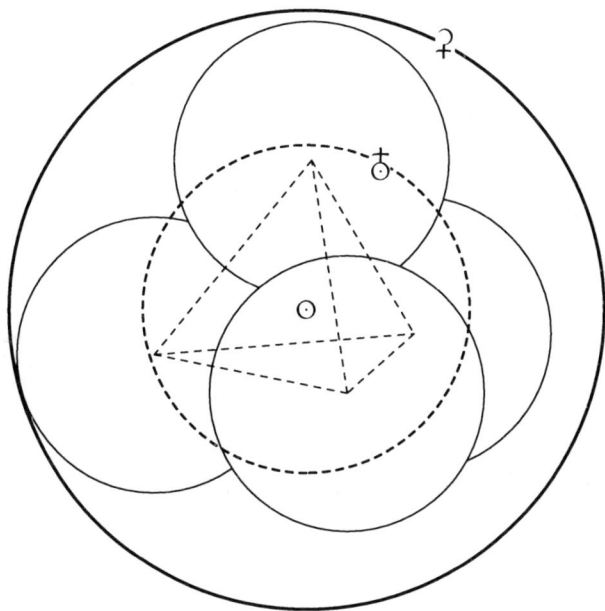

Mars and Jupiter

Mars and Jupiter are separated by the asteroid belt. To draw their relative orbits:

Draw a circle and then a square inside it. Next draw quarter circles from the corners of the square to the centres of its sides. Finally, draw a central circle which touches the four quarters.

If the outer circle is taken as Jupiter's mean orbit then the inner circle is Mars' mean orbit with over 99.9% accuracy.

Two other equivalent methods are shown below.

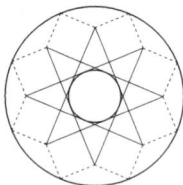

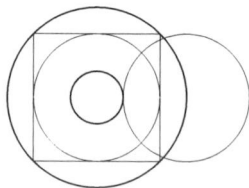

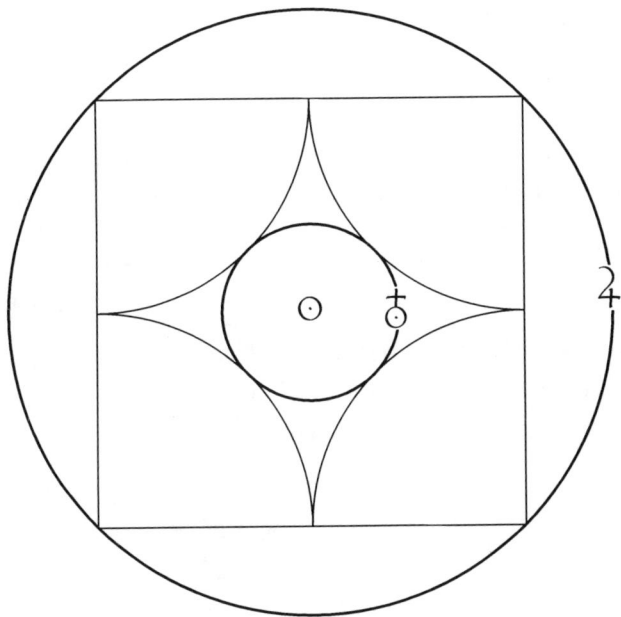

Jupiter and Saturn

Jupiter and Saturn are the two largest planets and the furthest two which can be seen with the naked eye.

If Jupiter's orbit round the Sun is represented by two Moon diameters then Saturn's orbit is the size of the Earth. This is 6:11 and well over 99.9% accurate.

For the keen - Jupiter's dance with Saturn is shown below left, and its experience of Uranus below right.

How fine - that the proportions of the orbits of the two slow giants of the sky should be the octave of our own home pair. The triangle is the rotary form of the octave.

The Sun and the Moon seem the same size from Earth.

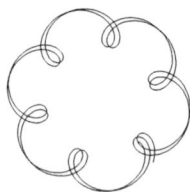

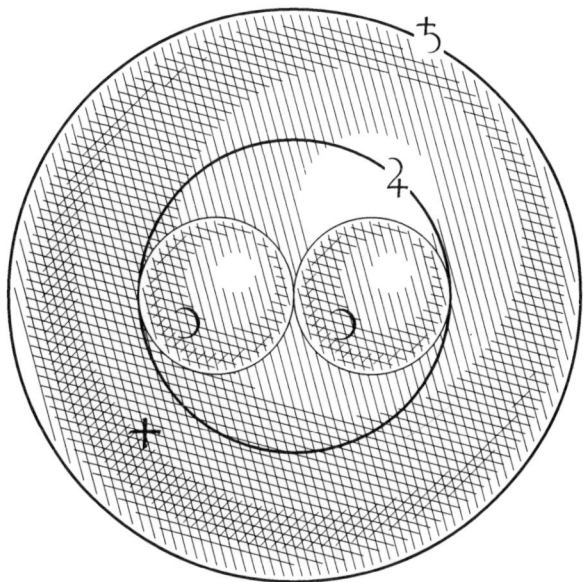

SATURN PIE

The circumference of Mars' orbit is the radius of Saturn's orbit and the circumference of Saturn's orbit is the diameter of Neptune's orbit.

So take a circle to represent Mars' mean orbit round the Sun. Unwind the circle into a line and draw a new circle (Saturn) using the line as a radius. Unwind the new circle into a straight line and with this as diameter make a final circle. It will be Neptune's orbit with 99.9% accuracy.

The relationship of a circle's circumference to its diameter (length round to length across) is known as pi, or π, and is best approximated as 22/7 (see page 6).

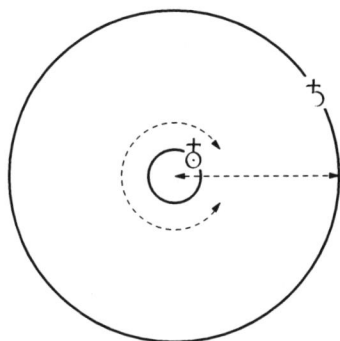

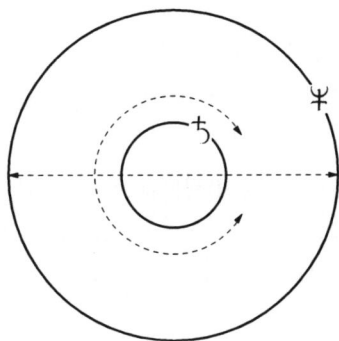

Saturn and Earth and Mercury

Here is a coincidence - or a tri-incidence to be precise:

There are only two occasions in the solar system when two planets' orbits mirror their sizes - and both involve our home planet Earth.

The innermost and outermost, or fastest and slowest, visible planets from Earth are Mercury and Saturn, and these are the other two planets involved.

Top left shows a thirty-pointed star defining Earth's tiny orbit from Saturn's and then, right, shows their relative sizes to be in the same proportion.

Bottom left shows a solution for Earth and Mercury's orbits (see page 10) which also gives their relative physical sizes, bottom right. This is very queer.

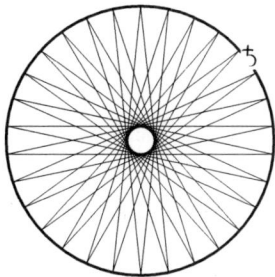

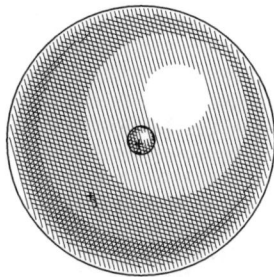

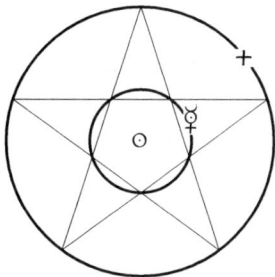

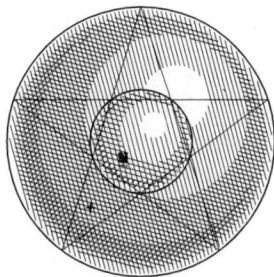

Saturn and Uranus

Saturn's orbit round the Sun has a radius almost exactly one half that of its neighbour Uranus. This proportion, shown opposite as an equilateral triangle, is 99.5% accurate.

With the discovery of Uranus in 1781, the ancient system of seven heavenly bodies began to become disrupted. Uranus was the first of the newly discovered planets which could be detected by means other than the clear unaided eye. This qualitative difference between the visible and invisible planets is evident here as Uranus' whole octave spacing from the furthest limit of the visible planetary system.

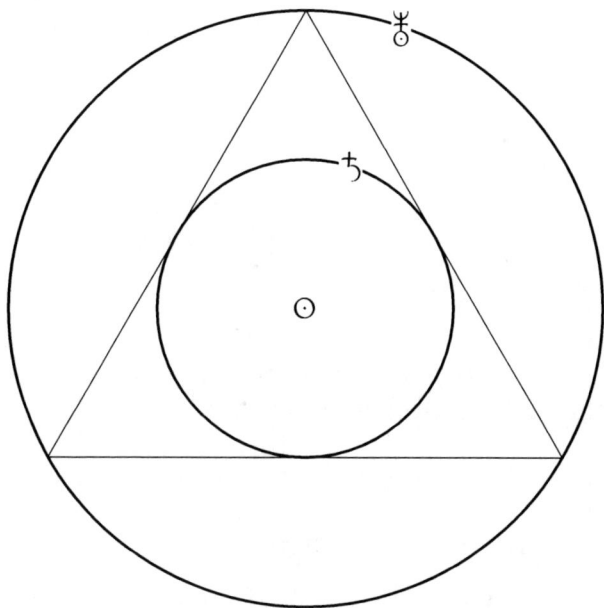

Saturn and Neptune

Imagine you are looking down on a tetrahedron made of touching spheres (see page31). You will see the four small circles shown opposite.

Draw a triangle to enclose the figure.

Draw a circle round the triangle.

If the size of one of the small circles is Saturn's orbit then the outer circle is Neptune's mean orbit with 99.9% accuracy.

In ancient astrology Saturn rules contraction, the opposite to Jupiter (expansion). It is too early for anyone to really know what Neptune is about; it hasn't even been round the heavens once since its discovery.

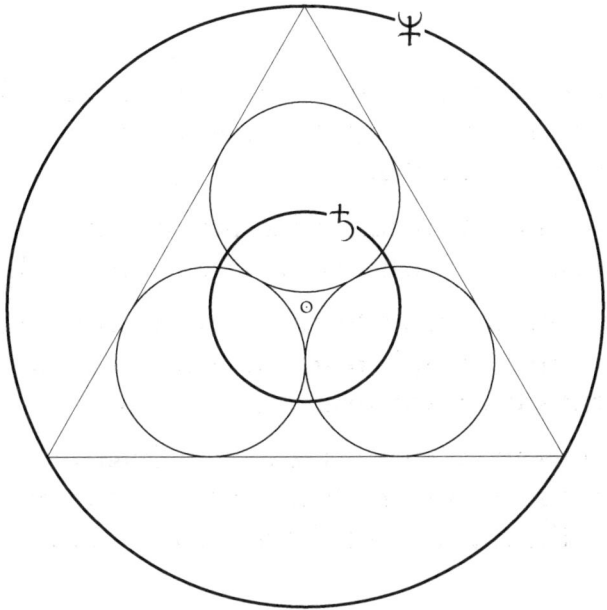

Uranus and Neptune

The tetrahedron is the most simple regular solid there is. It has four triangular faces and is the first of the five 'Platonic' solids.

Now cut the corners off to produce an equal-edged 'Archemedian' solid called a truncated tetrahedron.

If a sphere round a tetrahedron represents Neptune's orbit, then the sphere round its truncated self defines Uranus' orbit with over 99.9% accuracy.

Uranus and Neptune are the two huge outer planets of the solar system. Neither can be seen from Earth with the unaided eye. Their dance is shown below.

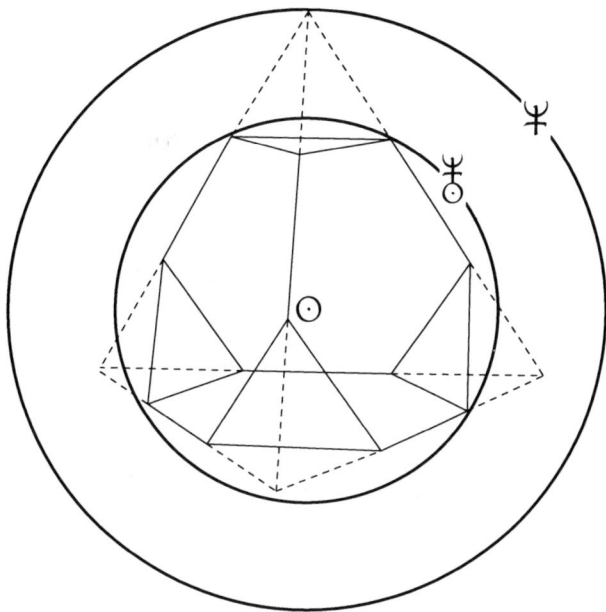

A Master Diagram

Long Meg Druid's Circle in Cumbria, England, is shown opposite. It is a type-B flattened stone-circle, a precise and common stone circle shape. North is up and the midwinter Sun sets over Meg in the southwest.

Three small diagrams show three different solar system relationships invoked from the circle centre. Left to right:

Mean orbits as page 11. Venus and Earth (99.7%).

Mean orbits standing on Earth. Nearest stone is the Sun, furthest stones pushed out by Mercury (99.9%).

Earth centred. Saturn nearest : Saturn furthest (99.9%).

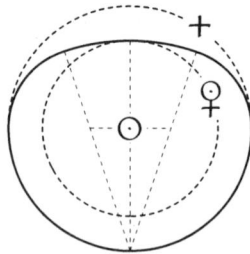

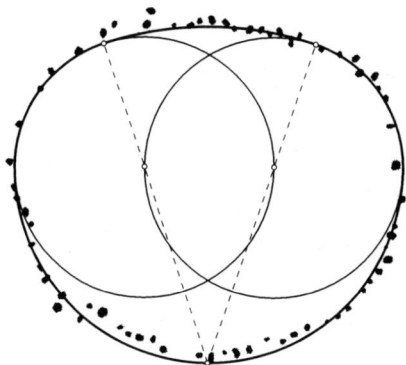

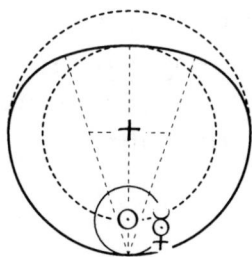

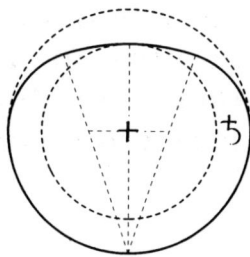

VIEWS FROM EARTH

The set of pictures opposite shows the movements of Mercury, Venus, Mars, Ceres, Jupiter and Saturn round Earth. The scales are all different. We are in the centre of each pattern (see page 16).

Mercury divides the heavens into three, kissing Earth every 116 days. Venus we have seen before - a perfect division of the heavens into five over eight years.

Mars is not so beautiful from Earth, coming close every 780 days to roughly divide the heavens into seven. The motions of Ceres divide the heavens precisely into eighteen, coming close every 467 days. She dances this pattern in exactly twenty-three years - to the day!

Jupiter makes eleven approaches over twelve years (five days off). Saturn does twenty-eight loops, coming close every 378 days, then fills the gaps to make a perfect division into fifty-seven (over fifty-nine years).

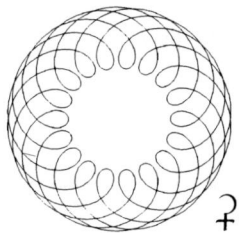

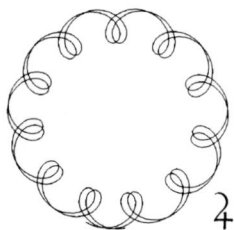

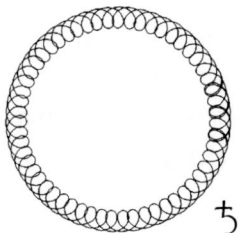

Bode's Law & Geddes' Oddity

Bode's law runs as follows: Take the series 0, 3, 6, 12, 24, 48, 96 and 384 - and add four to each.

This gives 4, 7, 10, 16, 28, 52, 100, 196, 388. Now, if the Earth's distance from the Sun is 10, then this series of numbers represents the orbits of the planets (missing Neptune) with passable accuracy.

Geddes' Oddity is between multiplied ideal orbits and, using abbreviations for the planets, runs as follows:

Ve.Ur = 1.204 Me.Ne   Ve.Ma = 2.872 Me.Ea
Me.Ne = 1.208 Ea. Sa   Sa.Ne = 2.876 Ju.Ur
Ea.Sa = 1.206 Ma.Ju   Ve.Ma.Ju.Ur = Me.Ea.Sa.Ne

Geddes' Oddity seems to show the outer four large planets as a mirror of the inner four small planets, reflected in the asteroid belt. The multiplications are shown diagrammatically opposite.

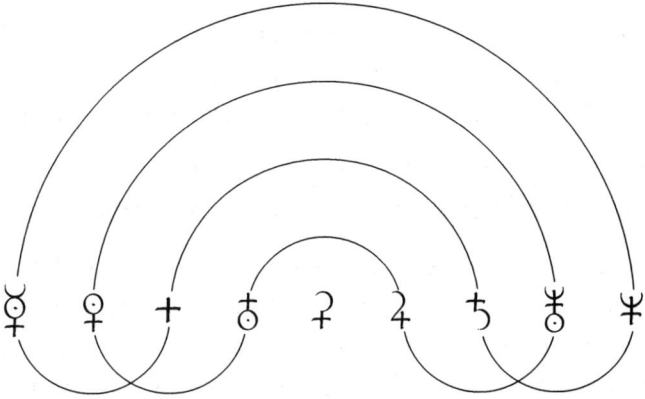

MUSIC

The ordinary periods of the planets produce a number of not very thrilling relationships. The key to the puzzle lies in their timings relative to each other.

The table opposite shows the time (in days) between pairs of visible planets' closest approaches, or kisses.

Amazingly, four Mercury-Venus kisses take the same number of days as five Mercury-Earth kisses. Four Venus-Earth kisses take the same time as three Mars-Earth kisses, or seven Venus-Mars kisses or nine Venus-Ceres kisses. Four Mercury-Earth kisses take the same time as five Mercury-Ceres kisses. Four Mars-Jupiter kisses take the same time as seven Earth-Ceres kisses.

These are all highly accurate and that's just the start of it! Simple ratios like 4:3, 5:4, 5:3 or 7:4 are the building blocks of music and rhythm. Next time you are bored see how many you can find!

	☿	♀	⊕	♁	♂	♃	♄
☿	-----	144.6	115.9	100.9	92.83	89.79	88.70
♀	144.6	-----	583.9	333.9	259.4	237.0	229.5
⊕	115.9	583.9	-----	779.9	466.7	398.9	378.1
♁	100.9	92.83	89.79	-----	1162	816.5	733.9
♂	92.83	259.4	466.7	1162	-----	2744	1991
♃	89.79	237.0	398.9	816.5	2744	-----	7252
♄	88.70	229.5	378.1	733.9	1991	7252	-----

The number of days between planetary kisses

NOTES